THE SILENT FLASH: FROM AWARENESS TO RESILIENCE

This publication is designed to provide accurate and authoritative information regarding the subject matter covered. It is provided with the understanding that the author and publisher are not engaged in rendering legal, financial, engineering, or other professional services. If expert assistance is required, the services of a competent professional should be sought.

The views and opinions expressed in this book are those of the author and do not necessarily reflect the official policy or position of any organization, employer, or affiliated entity.

All trademarks and registered trademarks appearing in this book are the property of their respective owners.

Published independently by Jeff Kohl

ISBN: 979-8-234-05129-5

Cover design by: Jeff Kohl
Editing by: Jeff Kohl

For more information, visit:
www.tieronerg.com

Table of Contents

THE SILENT FLASH: FROM AWARENESS TO RESILIENCE

Preface

This book was not written out of fear, speculation, or fascination with extreme scenarios. It was written out of experience and out of concern for how quietly modern society has traded resilience for efficiency in the pursuit of scale and optimization.

For more than three decades, I have worked across information technology domains spanning applications, data, infrastructure, and enterprise operations. I have held senior executive leadership roles, advised in regulated industries, and worked closely with compliance, legal, audit, finance, and business teams in environments where continuity is not optional. More recently, I have served as an independent consultant, helping organizations navigate increasingly complex and interconnected technology ecosystems.

It was in that context that I became responsible for leading the implementation of electromagnetic pulse (EMP) protection for financial-sector data centers - the first commercial implementations of their kind in the United States. What struck me most was not resistance, denial, or cost sensitivity. It was unfamiliarity. Even among highly sophisticated organizations that invest tens of millions of dollars in technology, cybersecurity, and operational protections, EMP risk was poorly understood, often misunderstood, and rarely evaluated alongside other enterprise risks.

Once the nature, scale, and potential impact of EMP were clearly explained, decisions to invest in protection were made quickly. The hesitation was not philosophical. It was informational. EMP simply was not part of the standard risk vocabulary.

That observation became the seed for this book and ultimately led me to establish TierOne Resiliency Group, focused on advancing awareness of electromagnetic risk and supporting organizations as they confront the challenges of assessment, preparedness, and resilience.

Electromagnetic risk is not new. The physics have been understood for decades. Natural electromagnetic events have occurred throughout history. Man-made electromagnetic effects have been studied, tested, and documented. What has changed dramatically is our dependence on tightly coupled electronic systems and layered digital infrastructure.

Modern society now depends on a tightly coupled ecosystem of electronics, networks, cloud platforms, data centers, and automated decision systems that must function in near-perfect coordination across organizational, geographic, and sector boundaries. Control is fragmented, dependencies are opaque, and almost none of these systems were designed to withstand electromagnetic disruption.

This book does not argue that catastrophe is inevitable. It does not predict when or how an electromagnetic event might occur. It does not suggest that organizations or individuals should retreat from modern technology. Instead, it asks a more grounded and practical question: **What happens when assumptions about continuity no longer hold?**

We have seen this pattern before. Societies often act decisively to mitigate risk only after disruption forces awareness, such as after financial crises, natural disasters, or acts of terrorism expose weaknesses that had long gone unaddressed. In each case, the lesson is learned at a cost. The question facing leaders today is whether electromagnetic risk will follow the same path, or whether awareness and preparation can precede impact.

This book is intended to help close that gap.

It examines electromagnetic risk not as a standalone threat, but as a stressor applied to an already optimized and increasingly fragile system. It explores natural and man-made electromagnetic events, the limits of traditional disaster recovery, the dynamics of progressive collapse, and the dependency chains that now define business operations, public services, and daily life. It looks across all sixteen critical infrastructure sectors and considers what resilience means in an era of cloud computing, zero-trust security, artificial intelligence, and global interdependence.

Most importantly, it reframes the conversation away from fear and toward responsibility.

Preparedness is not pessimism. It is stewardship. In an interconnected world, resilience is no longer about controlling every dependency, it is about understanding them honestly and strengthening what can be strengthened while there is still agency to do so.

My hope is that this book serves as a foundation for informed discussion, thoughtful leadership, and deliberate action. This is not because disruption is guaranteed, but because continuity deserves the same seriousness, rigor, and foresight applied to every other mission-critical risk.

Disclosure

In the interest of clarity and accessibility, the author utilized AI-assisted tools to help refine language and structure throughout this book. The ideas, analysis, conclusions, and professional judgments presented are the product of the author's experience, research, and independent judgment. Responsibility for the content rests solely with the author.

PART I — ORIGINS, PHYSICS, AND CONTEXT

Chapter 1 — The Accident That Changed Everything

How It Started

Electromagnetic pulse risk was not discovered because someone was looking for it. It was discovered because something went wrong. In July of 1962, the United States conducted a high-altitude nuclear test known as Starfish Prime. A nuclear device was detonated roughly 250 miles above the Pacific Ocean. At the time, the objective was narrowly scientific: to understand how nuclear detonations behaved in space and how radiation interacted with the upper atmosphere.

What happened next was unexpected. More than eight hundred miles away, in Hawaii, streetlights went out. Telephone systems failed. Electrical relays tripped. Alarms triggered without cause. Satellites in orbit, assets that had taken years to design and deploy, were damaged or lost entirely. There was no blast on the ground. No fire. No visible destruction. Critical systems failed.

This was the moment electromagnetic pulse stopped being theoretical. The explosion had produced an intense electromagnetic disturbance that traveled through the Earth's magnetic field and coupled directly into electrical and electronic systems. Infrastructure that had never been considered part of a nuclear test suddenly became collateral.

What was revealed that day was not simply a new weapons effect. It was a new category of vulnerability, one that would grow more dangerous with every technological advancement that followed.

At the time, the implications were largely confined to military and scientific circles. Society was still largely mechanical. Electronics existed, but they were not yet foundational. The world could absorb the lesson without changing. That window has closed.

The Silent Flash

From Starfish Prime to the Digital Age

When the Starfish Prime nuclear test revealed electromagnetic pulse effects in 1962, society was only beginning its transition into the electronic age. Computing at the time was centralized, limited, and largely isolated. Mainframe systems occupied controlled environments. Power grids were simpler. Most economic and social functions remained mechanical or manual.

Over the decades that followed, this balance shifted decisively.

The 1960s and 1970s saw the rise of large-scale commercial computing and electronic control systems in government, defense, and industry. The 1980s introduced personal computers, embedding electronics into offices and homes. The 1990s brought widespread software dependence and the global expansion of the internet, connecting systems that were once independent. The 2000s accelerated this trend through mobile computing, cloud services, and digitized financial markets. The 2010s introduced automation, robotics, and platform-scale data centers operating continuously at global scale.

Today, artificial intelligence constitutes a structural inflection point in national infrastructure. Core computational capacity, automated decision architectures, and industrial control systems are increasingly centralized within high-density electronic environments predicated on uninterrupted electrical power, thermal regulation, water supply, and network continuity. The projected water, energy, and material demands of AI-scale data centers are poised, within a single decade, to exceed the cumulative infrastructure expansion realized from the 1960s to the present, compressing generational growth into a period of extreme systemic concentration.

Over this sixty-year evolution, electronics shifted from being tools that supported society to systems upon which society depends, transforming EMP from a military curiosity into a systemic civilizational risk.

Experience Inside the System

Over the course of my career, I have worked inside the systems that quietly replaced mechanical resilience with electronic dependency, enabling data centers, financial platforms, networks, and control systems that underpin modern life. I have seen how efficiency, automation, and scale transformed entire industries, and how those gains steadily reduced margin for disruption, improvisation, and recovery. What was gained in speed and optimization was often lost in resilience.

As this book explores in detail, EMP is not merely a threat to the electric grid. It poses destructive risk to any system containing electronics and to the digital assets, data, and trust that depend on them. Business continuity, data integrity, and operational confidence all rely on electronic systems that were never designed to endure electromagnetic disruption. A common misunderstanding is that EMP risk is defined solely by large-scale, catastrophic events. Electromagnetic disruption spans a wide range of magnitudes, sources, and delivery mechanisms. Localized incidents can produce severe operational and financial harm. These smaller-scale events are not peripheral concerns; they are central to the modern electromagnetic threat landscape. Awareness and education are therefore foundational.

The Infrastructure We All Depend On

The Department of Homeland Security formally recognizes sixteen critical infrastructure sectors that together form the backbone of modern society, supporting the continuity of services relied upon by individuals, businesses, and governments alike. Every one of these sectors and the organizations that operate within them depends heavily on vulnerable electronics and is therefore exposed to electromagnetic risk. Each will be examined in detail later in this book.

Despite decades of study and documented evidence, the lack of awareness persists. When organizations are equipped with a clear understanding of the risk, they are positioned to make an informed choice: to pursue technical assessment and protection, or to consciously accept the exposure. What cannot be justified is ignorance by default.

Bottom Line

Starfish Prime was not a warning shot fired at modern society. It was a glimpse of a future we were not yet prepared to understand. The physics revealed in that moment never changed. What changed was our dependence on an ever-expanding array of electronic systems, systems now deeply embedded in every aspect of economic activity, governance, and daily life.

The danger of EMP does not lie in its novelty. It lies in how long we have known, and how much more vulnerable we have become.

Too often, societies take meaningful action only after disaster or material harm has already occurred. Financial crises led to reforms such as Sarbanes–Oxley. Earthquakes reshaped building codes. Flood disasters drove levees and land-use planning. Terrorism transformed aviation security and financial monitoring. In each case, awareness followed impact. Protection followed failure.

The open question is whether electromagnetic risk will follow the same reactive path or whether preparation will precede disruption.

The purpose of this book is to raise awareness, advance education, and accelerate a deliberate mobilization of effort, before continuity is tested, rather than after it is lost.

Chapter 2 — How EMP Works (When Technology Meets Physics)

Electromagnetic pulse is often discussed in extremes, either buried in dense physics or exaggerated into spectacle. Neither approach is helpful. At its core, EMP is not mysterious. It is a problem of energy overwhelming design limits.

Modern electronic systems operate within precise tolerances. Voltages are tightly controlled. Signals are finely timed. Components are optimized for efficiency rather than abuse. When a sudden electromagnetic surge is introduced, those assumptions collapse. Conductors behave like antennas. Energy couples into circuits faster than protective mechanisms can respond. In this sense, EMP is simply physics asserting itself over engineering assumptions.

Technological progress has unintentionally increased vulnerability. As systems have become faster, smaller, and more efficient, the margin for electromagnetic disturbance has narrowed. Lower operating voltages and higher clock speeds leave less tolerance for anomalies. In many cases, EMP does not need to melt hardware or leave visible damage to be dangerous. The most destabilizing outcome is often ambiguity, systems that appear intact but behave unpredictably. That uncertainty erodes trust, delays recovery, and compounds risk.

EMP is not a single phenomenon, but a sequence of electromagnetic effects that unfold over different time scales and interact with infrastructure in distinct ways. These effects are commonly categorized as E1, E2, and E3. Understanding each is essential, because modern systems are rarely exposed to only one phase. Meaningful resilience must account for all three.

The E1 pulse occurs first and is the fastest and most intense. Unfolding in nanoseconds, it directly threatens microelectronics and digital systems by coupling energy into circuits before protective devices can react. Traditional surge protection and circuit breakers are largely ineffective at this speed. E1 primarily affects microprocessors, servers, routers, switches, storage systems, industrial control equipment, sensors, medical devices, and communications hardware. It does not need to physically destroy components to be harmful. By corrupting logic states or damaging semiconductors, E1 undermines the reliability of systems that may still appear operational. In a society where nearly all coordination depends on digital electronics, E1 attacks the nervous system.

The E2 pulse follows on a slower timescale, unfolding over microseconds to milliseconds. Its effects resemble those produced by lightning strikes, and many systems are already designed with lightning protection in mind. On its own, E2 is often manageable. Its danger lies in timing. If E1 has already degraded protective electronics, E2 can exploit those weakened defenses. Systems that would normally withstand surge events may fail because upstream protection has been compromised. E2 affects power distribution equipment, communications lines, antennas, transportation signaling, and industrial control networks. It rarely acts alone, but it compounds damage in already stressed systems.

The E3 pulse is fundamentally different. Occurring over seconds to minutes, it interacts with Earth's magnetic field and induces currents in long conductors such as high-voltage power lines, pipelines, and undersea cables. E3 effects are similar to those produced by severe solar storms and geomagnetic disturbances. They can saturate transformers, overheat internal components, and cause permanent physical damage to grid infrastructure. E3 primarily threatens high-voltage transmission systems, large power transformers, and regional or national power grids. Unlike digital electronics, these assets cannot be rebooted or replaced quickly. Damage often results in recovery timelines measured in months or years, not days.

What makes EMP especially dangerous is that modern infrastructure is rarely vulnerable to only one of these effects. E1 disables control and coordination. E2 exploits weakened protection. E3 removes the power foundation everything depends on. Together, they create conditions in which failure is not only immediate, but also difficult, sometimes impossible to recover from on meaningful timelines.

EMP resilience is therefore not about defending against a single pulse. It is about engineering survivability across time scales, from nanoseconds to minutes, in systems that were never designed to endure such stress.

Chapter 3 — The Two Sources: Nature and Intent

EMP risk does not originate from a single adversary or scenario. It arises from two fundamentally different sources, each dangerous in its own way.

Natural Electromagnetic Events

Solar activity periodically releases enormous bursts of energy into space. When these events interact with Earth's magnetic field, they induce electrical currents in long conductors such as power lines, pipelines, and communication cables.

The Carrington Event of 1859 is often cited because it occurred before modern electronics existed and still managed to disrupt global telegraph systems. If a comparable event occurred today, it would interact simultaneously with satellites, GPS, power grids, financial networks, and communications systems. Nature offers no intent. No negotiation. No accommodation for readiness.

Man-Made EMP

Man-made EMP, nuclear and non-nuclear, is designed to exploit technological dependence. It bypasses conventional defenses, disabling systems without destroying infrastructure or causing immediate casualties.

What unites natural and man-made EMP is outcome, not motive. In both cases, modern systems are exposed in ways their designers never intended.

Chapter 4 — The Events We Rarely Discuss

Electromagnetic disruption is often treated as a hypothetical risk, something belonging to the future, or realm of unlikely catastrophe. History tells a different story. Electromagnetic effects have occurred across centuries and technologies, leaving documented evidence, but rarely synthesized into a broader understanding of systemic vulnerability.

In the nineteenth century, powerful geomagnetic storms disrupted global telegraph networks, inducing currents strong enough to shock operators and ignite equipment. During the Cold War, high-altitude nuclear testing generated electromagnetic effects damaging satellites, disabling streetlights, and interfered with communications hundreds of miles away. In the modern era, solar storms have repeatedly degraded GPS accuracy, disrupted radio communications and stressed power grids forcing utilities and satellite operators to take emergency protective actions. None of these required malicious intent to expose weakness; physics was sufficient.

What is rarely discussed is not the occurrence of these events, but how they are framed. They are labeled as space weather incidents, isolated engineering anomalies, or historical military artifacts, categorized in ways that prevent them from being recognized as warnings about the fragility of modern, electronically dependent systems. Each incident is treated as an exception rather than a signal. The result is a fragmented narrative where clear evidence of vulnerability exists, but responsibility for response remains diffuse.

The absence of sustained public attention does not indicate safety. It reflects the difficulty of assigning ownership across sectors, the challenge of justifying investment without visible damage, and the lack of a single, dramatic event that forces recognition. Risk does not always announce itself with spectacle. It accumulates quietly, in the space between what is known and what is acted upon.

That gap between demonstrated vulnerability and sustained, coordinated response is where electromagnetic risk resides.

PART II — WHEN SYSTEMS DEPEND ON EVERYTHING WORKING

Chapter 5 — Why Electronics Changed Everything

A Walk Through a "Simple" On-Premises Workday

Consider an employee arriving at the office to begin what appears to be a routine workday. The building has power. The lights are on. The employee badge-swipes into the facility and takes a seat at a company-managed workstation. From the outside, the organization appears self-contained, secure, and operational.

The employee attempts to access a core business application hosted inside the company's own data center. Although the user is physically on-premises, access still requires two-factor authentication. A password is entered, followed by a secondary authentication request generated by a cloud-based identity provider. That request leaves the building almost immediately, traversing internal switches, passing through firewalls, and routing across telecommunications infrastructure to an external identity service hosted in a third-party cloud provider's data center. That provider, in turn, depends on its own power, cooling, networking, and electronic security controls. Only after the identity service validates the request and returns an authorization response does access proceed. Before the employee reaches a single internal application, multiple organizations, data centers, and infrastructure sectors have already been engaged.

Even after authentication, the session is not trusted. Corporate policy requires inspection through a Zero Trust security platform. Although the employee is inside the building, their traffic is evaluated by a cloud-hosted security service that continuously enforces access policies, monitors behavior, and authorizes application connections in real time. Once again, traffic exits the building, crosses external networks, and is processed by infrastructure the organization does not own or operate. Security has improved, but dependency has increased.

The Silent Flash

The application is hosted in the company's on-premises data center, which relies on commercial power, backup generators, uninterruptible power supplies, cooling systems, and network connectivity. Inside the facility, servers, storage arrays, switches, and control electronics coordinate to process the request. The application does not function in isolation. To complete its task, it calls several external services: a public cloud platform, a business partner's on-premises system, and a managed SaaS provider. Once processing is complete, data is transmitted to a financial institution for settlement, reporting, or compliance.

From the employee's perspective, the experience feels local and seamless. From a systems perspective, it is anything but. A single login touches internal networks and on-premises infrastructure, cloud-based identity services, Zero Trust security platforms, telecommunications carriers, multiple data centers across providers, third-party services, and financial networks. Every step depends on power, cooling, electronic control systems, networks and coordination across organizations. Each dependency is necessary. Each dependency is electronic. Most lie beyond the organization's direct control.

This example illustrates why modern operational risk cannot be confined to the perimeter of a building or even a data center. Physical presence does not imply independence. Security improvements such as multi-factor authentication and Zero Trust architectures strengthen protection against cyber threats, but they also deepen reliance on external, electronically mediated services. In tightly coupled environments, disruption does not need to be total to be consequential. It only needs to interrupt coordination.

Electromagnetic disruption is uniquely positioned to stress this model because it targets the underlying electronics that enable identity, security, networking, and trust across the entire ecosystem. Understanding this dependency chain is essential, because resilience today is not about the survivability of a single system, but about the survivability of an interconnected, layered ecosystem.

Electronics changed that balance completely. Efficiency replaced redundancy. Automation replaced manual fallback. Centralization replaced decentralization. These changes delivered extraordinary gains, but they also removed buffers that once absorbed shock.

Chapter 6 — Cascading Failure

Immediate Impact vs. Progressive Collapse

One of the most dangerous misconceptions about electromagnetic pulse risk is the belief that it causes instant societal collapse. It does not. What EMP does instead is far more insidious: it strips away coordination, confidence, and predictability, the invisible qualities that allow complex systems to function.

When power fails, communications quickly follow. As communications degrade, emergency response falters. Healthcare outcomes worsen not because facilities disappear, but because systems cannot coordinate. Financial systems stall not because value vanishes, but because verification and trust do. In tightly coupled environments, even minor disruptions can cascade once visibility is lost. EMP accelerates this process dramatically because modern systems are interdependent by design.

This is why EMP risk cannot be evaluated in isolation. It is not merely about the failure of individual devices, but about the fragility of a layered ecosystem, systems that depend on simultaneous, coordinated operation across countless components and external infrastructure sectors, most of which lie beyond the control of any single organization.

Collapse, when it occurs, is rarely cinematic. It unfolds. Some impacts are immediate: loss of power, communications, or electronic control systems. Others emerge gradually over hours, days, or weeks as dependencies fail, and recovery assumptions unravel. This distinction matters. Early phases often appear manageable. Backup power activates. Manual procedures are invoked. Leaders assess damage and attempt coordination.

Over time, constraints surface. Fuel supplies tighten. Replacement components remain unavailable. Communications continue to degrade. Data integrity is questioned. Staff availability declines as personal hardships accumulate. What initially appeared to be a technical issue becomes operational. What seemed operational becomes systemic.

Progressive collapse is not a single failure, but the compounding interaction of many small ones. Operations slow rather than stop. Transactions queue but do not clear. Systems function intermittently and unreliably. Decision-making degrades as information becomes incomplete or untrustworthy. Most critically, dependencies assumed to be external, such as power, communications, logistics, cloud services and financial networks remain unavailable far longer than expected.

Business continuity plans often assume these services will be restored quickly or remain partially available. Progressive collapse invalidates those assumptions. In some scenarios, restoration timelines stretch from months into years. The result is not merely downtime, but erosion of confidence in systems, in data, and in recovery itself.

For consumers and communities, progressive collapse is not experienced as a single dramatic event, but as a series of accumulating disruptions. Access to cash becomes uncertain. Electronic payments fail intermittently. Fuel availability tightens. Medical services operate in constrained modes. Communications remain unreliable. Daily life grows more difficult not because everything has stopped, but because nothing works consistently. Uncertainty itself becomes destabilizing.

Case Study 1 — When Redundancy Depends on the Same Assumptions

During an enterprise resilience assessment of a multi-site data center environment supporting regulated financial and commercial workloads, leadership expressed strong confidence in the organization's continuity posture. The architecture appeared robust: geographically separated facilities, cloud-based failover capabilities, diverse network providers, and well-documented disaster recovery procedures that had performed reliably during routine outages. By conventional measures, the organization met and, in some cases, exceeded industry best practices for availability and continuity.

That confidence rested on a set of assumptions common across modern resilience models. Geographic separation was equated with independence.

Cloud failover was assumed to reduce exposure to physical infrastructure risk. Vendor diversity was believed to eliminate single points of failure. Recovery was expected to be coordinated through functioning communications and stable power. EMP risk was not explicitly dismissed; it was simply assumed to be implicitly addressed by existing continuity measures.

A deeper analysis shifted focus away from logical architecture and toward physical and electromagnetic dependencies. This examination revealed that, beneath layers of apparent redundancy, critical assumptions were shared across all sites. Facilities relied on the same regional power grids and transformer classes. Telecommunications diversity masked dependence on common carrier interconnection points. Power conditioning and control electronics were standardized across locations. Cloud regions, while logically distinct, remained physically exposed to shared grid segments and network infrastructure. In effect, redundancy existed, but independence did not.

Under these conditions, an electromagnetic event affecting shared dependencies could plausibly degrade or disable multiple sites simultaneously. Both primary operations and recovery coordination could be compromised at once, leaving no viable path to failover. The discussion shifted from a familiar question. *Can we fail over?* to a more fundamental one: *What assumptions must remain true for failover to be possible at all?* In an electromagnetic disruption, the assumptions of stable power, intact electronics, and functioning communications could not be guaranteed.

The lesson was not technical, but conceptual. Redundancy improves availability; independence determines survivability. Resilience models optimized for uptime can fail catastrophically when exposed to correlated physical risks. EMP did not reveal a gap in architecture, it exposed a gap in thinking.

Chapter 7 — Why Disaster Recovery Alone Is Not Enough

Disaster recovery is built on assumptions that do not hold under electromagnetic disruption. Traditional recovery strategies presume that failures are localized, sequential, and ultimately recoverable. They assume systems can be restored from backups, alternate sites can be activated, and external dependencies on power, communications, timing, and transportation will remain intact long enough to coordinate response. EMP challenges every one of these assumptions simultaneously. Depending on the type of event, electromagnetic effects can disable primary and backup systems at the same time, damage or destroy electronics rather than merely interrupt data, and disrupt power, network connectivity, and timing across wide geographic areas. High-altitude or geomagnetic events can overwhelm geographically separated recovery sites, while localized EMP exploits can target shared dependencies, such as power conditioning, network infrastructure, or control electronics that disaster recovery plans implicitly rely upon. In these scenarios, recovery is not delayed; it is prevented. Disaster recovery is designed to restore systems after failure. Electromagnetic resilience must prevent catastrophic failure in the first place. Without survivability engineered into infrastructure, recovery plans risk becoming theoretical exercises disconnected from physical reality.

If disaster recovery cannot be relied upon, resilience must be redefined. The next chapters examine how organizations misclassify EMP risk, often assuming that cybersecurity, cloud adoption, or regulation have already solved the problem. They have not.

PART III — THE ILLUSION OF COVERAGE

Chapter 8 — The Myth of "Cyber Covers This"

Over the last two decades, cybersecurity has become the dominant lens through which organizations assess existential risk. Boards receive regular briefings, budgets continue to grow, and maturity models advance year after year. This focus is not misplaced, but it is incomplete. Cybersecurity addresses adversarial intent: who is trying to gain access, manipulate systems, or steal data. Electromagnetic pulse addresses something fundamentally different: physics. Firewalls, encryption, identity controls, and zero-trust architectures all assume the continued operation of electronic systems. EMP bypasses these controls entirely. It does not exploit software vulnerabilities or configuration weaknesses; it overwhelms the physical limits of electronic design.

This distinction is not academic, it is foundational. Cybersecurity operates within an intact electronic environment. EMP attacks that environment itself. The confusion persists because modern systems feel more resilient precisely because they are more advanced. In practice, the opposite is often true. Today's infrastructure is optimized for speed, efficiency, density, and automation, not for exposure to extreme electromagnetic energy. Lower operating voltages, higher clock speeds, tighter tolerances, and dense integration reduce design margin and increase sensitivity. Cyber controls do not mitigate this reality; they depend on it.

The False Comfort of "Defense in Depth"

Cybersecurity frameworks emphasize layered defenses, an approach that has proven effective against software-based threats, intrusions, and adversarial misuse of digital systems. Against electromagnetic pulse, however, those layers often collapse simultaneously. Multiple security controls may share the same power source, rely on the same timing and synchronization systems, or operate within the same unprotected electronic environment. EMP does not defeat these controls one by one; it defeats the underlying assumption that they will be available at all when they are needed. Firewalls, identity systems, intrusion detection, and encryption are rendered irrelevant if the electronics that host them cannot function.

For this reason, cyber resilience and electromagnetic resilience must be understood as complementary, not substitutive disciplines. One cannot stand in for the other. Strong cybersecurity does not mitigate electromagnetic vulnerability, just as hardened infrastructure does not prevent cyber intrusion. Confusing the two creates a dangerous illusion of preparedness, offering confidence precisely where fragility remains. EMP is not a cyber problem. It is a system survivability problem, one that requires engineering for physical resilience alongside digital defense.

Chapter 9 — The Cloud Fallacy: Risk Outsourced, Not Eliminated

A frequent response to discussions of electromagnetic pulse risk is a simple one: *"We're in the cloud."* The implication is that responsibility for extreme physical risk has been transferred to someone else. It has not. Cloud adoption does not eliminate EMP exposure, it relocates it. Cloud service providers operate vast physical infrastructures that include data centers, power substations, network interconnects, terrestrial and undersea cables, and edge facilities. These assets remain governed by the same electromagnetic laws as any on-premises environment. Cloud architectures are engineered for availability, not survivability; they assume functioning power grids, intact networking equipment, and recoverable electronics. EMP challenges all three simultaneously.

Cloud computing also concentrates dependency. What appears as geographic redundancy in logical architecture diagrams often relies on shared grid segments, common transformer classes, regional network hubs, and standardized hardware deployed at scale. EMP does not respect ownership boundaries. It affects electronics regardless of who operates them. When cloud services fail, accountability does not disappear, it returns to leadership, regulators, and customers who assumed resilience without verification. Service-level agreements focus on uptime under normal operating conditions; they do not guarantee survivability against electromagnetic pulse, geomagnetic storms, or nuclear effects, and force majeure exclusions are explicit. Customers therefore retain responsibility for understanding and managing systemic risk, even when workloads are outsourced. Cloud migration shifts operational burden, but it does not absolve fiduciary responsibility. Cloud services are tools, not shields. They improve efficiency, scalability, and flexibility, but they do not change physics, eliminate dependency, or remove leadership accountability. In an EMP scenario, the question will not be where systems were hosted, it will be why known risks were assumed away.

PART IV — ACTORS, EVENTS, ASYMMETRIC RISK

Chapter 10 — Nation-States and Strategic Doctrine

EMP does not exist on the fringe of strategic thinking. It is explicitly acknowledged often quietly, sometimes obliquely, in the military doctrine, strategic writings, and capability development of multiple nation-states. For adversaries confronting technologically advanced societies, electromagnetic disruption offers a form of asymmetric leverage that is difficult to ignore. It enables disruption without occupation, paralysis without invasion, and strategic impact without the immediate attribution that traditionally accompanies kinetic attack.

Several nation-states have explicitly acknowledged, studied, or incorporated electromagnetic pulse effects into military doctrine and capability development, demonstrating that EMP is neither speculative nor confined to a single adversary.

China has publicly and doctrinally emphasized electromagnetic and information warfare as a means of paralyzing technologically advanced opponents, with open-source analyses indicating sustained investment in both high-altitude nuclear EMP concepts and non-nuclear electromagnetic and high-power microwave systems as part of broader "systems destruction warfare."[1]

Russia has long articulated interest in so-called "new physical principles" weapons, a category that includes directed-energy and electromagnetic effects, and Western defense assessments consistently identify Russian research and operational planning related to EMP-relevant capabilities alongside strategic and non-strategic nuclear forces.[2]

North Korea has been repeatedly cited in U.S. congressional testimony and defense analyses as possessing nuclear weapons and ballistic missile systems that could be employed in a high-altitude detonation scenario capable of producing EMP effects over regional or continental scales. [3]

Iran has similarly appeared in multiple intelligence and policy assessments as pursuing missile capabilities that, if paired with a nuclear device, could support an EMP attack profile, particularly given Iranian doctrinal interest in asymmetric disruption rather than direct confrontation. [4]

Beyond major powers, open-source reporting suggests that **additional states and actors** are exploring or acquiring limited electromagnetic or high-power microwave technologies, lowering barriers to entry and expanding the range of potential threat actors. [5]

The **United States** itself has developed and tested non-nuclear electromagnetic systems, such as the Counter-electronics High Power Microwave Advanced Missile Project (CHAMP), underscoring that electromagnetic effects are treated as a legitimate and operationally relevant domain of modern warfare.[6] Collectively, these developments illustrate that EMP capability is not monolithic, rare, or theoretical; it is distributed across multiple actors, spans nuclear and non-nuclear systems, and is increasingly aligned with doctrines that exploit dependency, ambiguity, and cascading infrastructure failure rather than overt destruction.

In this context, civilian infrastructure becomes the target not because it is undefended, but because it is indispensable. Financial systems, communications networks, energy grids, transportation corridors, and data infrastructure are attractive precisely because their failure cascades rapidly into social, economic, and political instability. These systems are not merely support functions; they are the operating environment of modern life. Disrupting them undermines confidence, coordination, and continuity simultaneously.

EMP aligns with modern strategic doctrine because it exploits dependency rather than confrontation. It targets the assumptions embedded in highly optimized systems such as continuous power, persistent connectivity, synchronized timing, and electronic trust, rather than physical assets alone. The objective is not destruction for its own sake, but disruption that compounds faster than response.

Importantly, nation-state EMP capability is not monolithic. No serious actor relies on a single weapon or delivery model. Instead, capabilities are developed as portfolios designed to support a range of strategic objectives, from deterrence and signaling to regional disruption and coercion. These portfolios span a broad spectrum, including strategic high-altitude nuclear systems capable of wide-area electromagnetic effects; regional or theater-level delivery platforms; non-nuclear electromagnetic weapons intended for localized disruption; and directed-energy or high-power microwave systems optimized for targeting electronics and control nodes. This diversity matters because it means electromagnetic effects are not confined to all-or-nothing scenarios. They can be applied selectively, incrementally, and ambiguously, below traditional thresholds of escalation.

Access, more than geography, is the enabling factor. Modern infrastructure is globally interconnected by design. Satellite systems, global telecommunications networks, shared cloud ecosystems, and internationally distributed supply chains create pathways that transcend borders. Distance is no longer a meaningful barrier. Electromagnetic risk today is shaped less by where an adversary is located and more by the density and concentration of dependency within the target environment.

Strategic ambiguity further enhances the attractiveness of EMP capabilities, particularly non-nuclear and localized forms. Electromagnetic effects can be difficult to attribute with confidence, especially when they manifest as equipment failures, network instability, or unexplained outages rather than visible destruction. This ambiguity aligns with doctrines that emphasize escalation control, deniability, and the ability to test responses without committing to open conflict. EMP can degrade confidence without overt destruction, disrupt coordination rather than infrastructure alone, and complicate both technical diagnosis and political response.

For organizations, the relevance of nation-state capability is not about predicting intent or assigning blame. It is about recognizing that the tools exist, access exists, and systems are exposed. Risk does not require constant hostility. It requires only that capability and opportunity intersect during periods of stress, whether geopolitical tension, natural disruption, or systemic overload.

This is why electromagnetic resilience cannot be dismissed as speculative. It sits at the intersection of known physics, documented capability, and increasing systemic fragility. Preparedness, in this context, is not reactionary. It is prudence.

PART V — WHEN FAILURE IS CATASTROPHIC

Chapter 11 — Non-Nuclear EMP and Asymmetric Risk

EMP capability is no longer confined to nuclear powers or nation-state arsenals. Advances in electronics, energy storage, and directed-energy technologies have enabled non-nuclear electromagnetic devices capable of disabling localized, but highly critical infrastructure. These systems cannot produce continental-scale effects, but they do not need to. Modern society depends on concentrated nodes of control and coordination. A single data center, a telecommunications switching facility, or a control room can serve as a linchpin for entire regions or industries.

When one of these nodes fails, the impact rarely remains local. Dependencies propagate outward through networks, supply chains, and digital services, creating cascading effects far beyond the original point of disruption. Systems designed for efficiency and centralization amplify this fragility. What appears to be a contained incident quickly becomes a multi-sector problem as coordination, verification, and control degrade simultaneously.

Non-nuclear EMP devices lower the barriers to entry dramatically. They are easier to conceal, easier to deploy, and far more difficult to attribute with certainty. Their use blurs the line between sabotage, terrorism, criminal disruption, and warfare. Attribution delays complicate response, while ambiguity constrains escalation and accountability. As the underlying technologies diffuse, so too does the risk. EMP is no longer solely a strategic, nation-state concern, it has become an asymmetric one.

The most dangerous misconception surrounding EMP is the belief that only catastrophic, large-scale events matter. Scale is not the determining factor. Small, localized electromagnetic events can trigger disproportionate consequences when they target coordination nodes rather than physical assets. Disrupting identity systems, network control points, or industrial controllers often produces far greater impact than damaging infrastructure outright. Low-cost capabilities can disrupt systems worth billions of dollars, particularly when resilience has not been intentionally engineered.

EMP risk, therefore, is not defined by the size of the event, but by where it strikes and what it disconnects. In highly optimized, tightly coupled systems, modest disruptions can produce outsized outcomes turning localized electromagnetic events into systemic failures.

Chapter 12 — Solar Storms: The Threat Without Intent

Unlike man-made EMP, solar storms require no adversary. They are an unavoidable consequence of stellar physics. When coronal mass ejections interact with Earth's magnetic field, they induce powerful electrical currents in long conductors such as power lines, pipelines, and communication cables. These induced currents place extraordinary stress on systems that were never designed to absorb them.

Modern infrastructure is particularly exposed. Power grids, satellites, aviation systems, GPS, financial networks, and global logistics platforms all rely on electronics, timing, and long-distance transmission. A severe solar storm would not affect these systems sequentially, it would stress them simultaneously. The Carrington Event of 1859 occurred in a world without modern electronics, yet it still caused global disruption by disabling telegraph systems across continents. Today, a comparable event would interact with nearly every critical infrastructure sector at once. Modern society has never experienced a Carrington-scale event under today's level of electronic dependence.

Solar storms do not wait for preparedness. They do not respond to deterrence or diplomacy. They arrive when physics dictates, not when systems are ready.

While space weather forecasting can sometimes provide hours or even days of warning, prediction is not protection. Many vulnerable systems cannot be shut down safely without causing damage. Others cannot be disconnected in time, or cannot be restored quickly once affected. In practice, warning windows often collide with operational realities that limit meaningful mitigation.

In this context, preparedness matters far more than prediction. Forecasts may signal when stress is coming, but only resilience determines how systems endure it.

Chapter 13 — What Fails First

EMP does not affect all systems equally. The earliest failures occur in microelectronics - servers, routers, switches, industrial control systems, embedded controllers, and sensors. These components form the nervous system of modern infrastructure, quietly coordinating power, communications, safety, and decision-making across every sector.

When these systems fail, visibility disappears. Control rooms go dark. Dashboards freeze. Automated processes stop responding. Operators lose situational awareness before they fully understand what has happened or why. Data may still exist, equipment may still be standing, but the ability to observe, verify, and control conditions is gone.

EMP's first impact is not chaos. It is silence.

That silence is profoundly destabilizing. In complex systems, the absence of feedback is more dangerous than obvious destruction. Alarms do not trigger. Status indicators do not update. Commands are issued but nothing responds. Decision-makers are forced to act without reliable information, turning routine operations into guesswork.

Critically, this loss of control precedes any loss of physical structure. Buildings remain intact. Equipment appears undamaged. Infrastructure still stands, but it is no longer governed. In modern systems, loss of control is more destabilizing than visible destruction. Without control, safety systems cannot be verified, processes cannot be coordinated, and response degrades rapidly. This is the moment where cascading failure begins, not with collapse, but with uncertainty.

Chapter 14 — Why Recovery Is Not Guaranteed

Most recovery plans are written around a single, critical assumption: that the world surrounding the organization will continue to function. They presume transportation networks will remain operational, fuel will be delivered on schedule, replacement parts can be manufactured and shipped, and skilled personnel will be able to coordinate across organizations and jurisdictions. These assumptions are reasonable in the context of localized disasters. They are deeply flawed in the context of electromagnetic disruption.

An EMP event challenges every one of these assumptions at the same time. In a widespread electromagnetic scenario, the very systems required to enable recovery are among the most vulnerable. Transportation depends on powered infrastructure. Fuel distribution relies on electronically controlled pumping and logistics systems. Manufacturing is driven by automation and just-in-time coordination. Organizational response depends on communications, timing, and shared situational awareness. Even the most well-prepared organizations discover that their recovery strategies rely on dependencies far outside their direct control.

Traditional disaster recovery models assume failure is localized, sequential, and ultimately recoverable. EMP introduces simultaneity. Primary and backup systems may be affected at the same time. Geographically separated sites can share the same failure modes. Control electronics may fail across regions simultaneously, eliminating the ability to coordinate restoration. This is not a matter of delay. It is a condition that prevents recovery from initiating at all.

Recovery, in this context, is not guaranteed. It is conditional. It depends on intact infrastructure, available labor, functioning markets, and coordinated governance. Electromagnetic disruption threatens all of these concurrently. When the environment required for recovery does not exist, continuity plans cease to be operational tools and become theoretical exercises, documents that describe intent rather than capability.

This reality shifts the emphasis from response to prevention. Disaster recovery restores systems after failure; electromagnetic resilience must prevent catastrophic failure in the first place. Increasingly, electromagnetic resilience differentiates leaders from laggards. Organizations that invest in protecting critical infrastructure and data preserve the option to survive disruption and recover faster than peers who rely solely on assumed continuity. During a material EMP event, disruption is unavoidable. However, organizations that have preserved the integrity of their systems are positioned to resume operations when power, communications, and external services return. Those that have not may find their infrastructure damaged, their data compromised, and their recovery timelines unworkable - resulting in loss of customers, erosion of market position, or business failure altogether.

Resilience determines who returns to the market, and who does not. Without survivability engineered into infrastructure, recovery plans risk becoming symbolic rather than functional.

If recovery cannot be assumed, the conversation must shift again. The next question is not technical. It is systemic. What happens when failure propagates across infrastructure sectors, markets, and society itself?

Case Study 2 — When Recovery Timelines Break Down

During a cross-sector resilience review focused on energy-dependent infrastructure, planners evaluated prolonged outage scenarios using conventional restoration models. Mutual aid agreements, shared spare equipment pools, and established transformer replacement strategies were cited as evidence that even extended disruptions could be managed. The prevailing view was that while the consequences of a large outage would be severe, recovery would ultimately follow familiar, well-documented processes.

Those assumptions mirrored the models used for storms, earthquakes, and other localized disasters. They presumed that damaged transformers could be replaced within weeks or months, that manufacturing capacity would scale in response to demand, that transportation and logistics networks would remain operational, and that skilled labor could be mobilized and coordinated across regions. On paper, the system appeared strained but survivable.

Closer analysis fundamentally altered that conclusion. Large power transformers were revealed not to be interchangeable or mass-produced assets, but custom-built components with manufacturing timelines that routinely ranged from twelve to twenty-four months and often longer. Global supply chains for these components were limited, concentrated, and shared across regions, meaning that surge capacity could not be assumed. Even more critically, the transportation, installation, and testing of replacement equipment depended on powered infrastructure, functioning communications, and coordinated logistics, the very systems most likely to be impaired in a widespread electromagnetic event.

In such a scenario, recovery assumptions collapsed. Even where spare equipment existed, the ability to move, install, and commission it depended on systems exposed to the same disruption. Recovery was no longer delayed, it was constrained. The review reframed resilience from a recovery problem into a preservation problem. If critical assets could not be replaced on timelines that sustained modern society, then preventing their damage became far more valuable than planning to replace them after the fact.

The implications extended beyond engineering. Many of the components underpinning critical infrastructure, including transformers, power electronics, control systems, and specialized equipment are sourced from overseas suppliers. In an electromagnetic disruption scenario, this dependency becomes a strategic vulnerability. Replacement timelines stretch from months into years when global supply chains are constrained, inaccessible, or politically disrupted.

Strengthening domestic manufacturing capacity therefore emerged not as an economic preference, but as a resilience imperative. Producing critical infrastructure components within the United States reduces exposure to geopolitical risk, shortens recovery timelines, and improves the nation's ability to respond to large-scale electromagnetic damage. Current efforts to encourage domestic manufacturing investment represent an important step, but they must be aligned explicitly with resilience objectives. Incentivizing the domestic production of high-impact, long-lead-time infrastructure components should be treated as a strategic pillar of EMP risk mitigation.

The core lesson was stark. Recovery plans fail when they assume the world around them still works. In a national-scale electromagnetic disruption, recovery is limited not by intent, but by capacity. Domestic manufacturing determines how quickly continuity can be restored. Electromagnetic resilience, at its core, is about protecting assets whose loss cannot be reversed on human or economic timelines.

PART VI — THE SCALE OF DEPENDENCY

Chapter 15 — Critical Infrastructure and Systemic Exposure

Why Sector-Level Risk Cannot Be Viewed in Isolation

Modern society functions not because individual systems are resilient, but because many fragile systems operate in coordination. Power enables communications. Communications enable finance. Finance sustains supply chains. Healthcare, water, transportation, and emergency services all rely on this same electronic foundation. When these systems work together, society appears stable. When coordination is lost, even partially, the consequences cascade rapidly.

Electromagnetic pulse risk exposes this fragility more clearly than almost any other threat. EMP does not respect organizational boundaries, ownership models, or sector classifications. It disrupts the electronic and electrical systems that every critical infrastructure sector depends upon, often simultaneously and without warning. Some sectors experience immediate failure; others fail because upstream services they rely on are no longer available or trustworthy.

The following sections examine each of the sixteen sectors through an electromagnetic risk lens, first individually, and then as part of an interconnected system. The goal is not to rank vulnerabilities in isolation, but to show how disruption propagates, where dependencies concentrate risk, and why resilience must be approached as a system-level responsibility rather than a collection of sector-specific fixes.

EMP risk is not theoretical. It is structural. And understanding where and how society is exposed is the first step toward meaningful resilience.

Critical Infrastructure Sectors

The United States formally recognizes sixteen critical infrastructure sectors whose disruption would have a debilitating impact on national security, economic stability, public health, or public safety. These sectors form the operating backbone of modern society. They are deeply interconnected, increasingly digitized, and overwhelmingly dependent on electronics and continuous power. Their resilience is not defined by any single system, but by the coordinated functioning of millions of electronic components operating across organizations, jurisdictions, and industries.

Electromagnetic pulse does not affect these sectors equally. Some are immediately and directly exposed, while others fail through cascading dependency as upstream services degrade or disappear. What unites them is a shared reliance on electronic systems that were never designed to operate in a high-electromagnetic-threat environment. Understanding EMP risk therefore requires examining each sector on its own terms and then understanding how failure in one propagates across the rest.

The **energy sector** represents the most extreme point of exposure because it underpins every other sector. Electric power generation, transmission, and distribution rely on high-voltage transformers, long transmission lines, and electronic control systems that are uniquely susceptible to electromagnetic disruption. Damage at this level cannot be repaired quickly; replacement timelines for large transformers are measured in months or years, not days. When energy fails, no other sector can operate normally for long.

Closely coupled to energy is the **communications sector**, which also faces extreme risk. Cellular networks, fiber amplification systems, satellite links, and emergency communications depend on both continuous power and precise electronic timing. EMP does not need to destroy towers to silence networks. It disrupts the electronics that coordinate traffic, synchronize systems, and enable routing. Loss of communications rapidly degrades emergency response and undermines recovery coordination across all sectors.

The **financial services sector** faces very high exposure because it depends not just on availability, but on trust. Financial systems require synchronized clocks, continuous data integrity, and uninterrupted communications to verify transactions and settle obligations. EMP threatens these foundations directly. Money does not disappear, but verification does, and without verification, markets stall and confidence erodes faster than capital can absorb.

The **information technology sector**, including data centers and cloud platforms, also faces very high risk. These environments concentrate vast amounts of sensitive electronics into dense facilities optimized for efficiency and availability. While logical redundancy is often emphasized, it frequently masks shared physical dependencies such as power sources, network hubs, and standardized hardware. EMP exposes the difference between availability and survivability.

In the **healthcare and public health sector**, the consequences of electromagnetic disruption translate directly into human impact. Diagnostics, imaging systems, electronic health records, pharmacy automation, and life-support equipment all depend on electronics. Extended outages do not merely inconvenience operations, they increase morbidity and mortality as care degrades or becomes unavailable.

The **water and wastewater sector** is highly exposed through its reliance on electronically controlled pumping stations and treatment facilities. When control systems fail, sanitation and potable water availability quickly become public health crises. Similarly, the **transportation sector**, including aviation, rail, maritime, and roadway systems, relies heavily on electronic navigation, signaling, and control. Disruption here impairs movement of people, goods, fuel, and emergency resources.

Emergency services face high risk precisely when demand surges. Dispatch systems, communications platforms, and coordination tools degrade under electromagnetic stress, undermining response when it is most needed. **Government facilities** mirror these vulnerabilities, as civilian infrastructure often lacks hardened continuity capabilities outside a limited number of sites.

The **defense industrial base** occupies a medium-to-high risk category. While certain systems are hardened, the broader supply chains, supporting infrastructure, and commercial dependencies that sustain defense production remain exposed. The **chemical sector** similarly faces medium-to-high risk, as industrial control system failures introduce safety, environmental, and public health hazards.

In the **nuclear reactors, materials, and waste sector**, safety systems are generally hardened, but reliance on offsite power, logistics, and electronic coordination remains a vulnerability. **Food and agriculture** systems may continue production initially, but refrigeration, processing, and distribution fail rapidly without power and communications. **Dams** may retain structural integrity, but loss of electronic monitoring and control increases downstream risk. **Commercial facilities**, including retail and service industries, depend on electronic payments, access control, and logistics, while **critical manufacturing** faces amplified disruption due to automation and just-in-time supply chains.

Across all sixteen sectors, the pattern is consistent. EMP risk is not isolated, and it is not linear. Failure in one sector accelerates failure in others, compressing response time and overwhelming recovery assumptions. The risk is not additive. It is multiplicative and it is structural.

The Systemic Reality

No critical infrastructure sector exists in isolation. Each is bound to the others through implicit layers of electronic dependency, shared assumptions, and continuous coordination. When energy fails, communications degrade or collapse, removing the ability to coordinate response or recovery. When communications fail, emergency services are paralyzed, unable to dispatch, synchronize, or operate at scale. Financial disruption follows quickly, not because value disappears, but because trust, verification, and settlement depend on functioning networks and synchronized systems. As financial systems falter, supply chains stall, payrolls are delayed, and confidence erodes. Healthcare failures then become inevitable, not as a secondary concern, but as a direct consequence of lost power, communications, data access, and logistics. What begins as a disruption in one sector rapidly propagates across all others.

This is why electromagnetic risk cannot be evaluated sector by sector or mitigated through isolated improvements. EMP exposure compounds across interdependent systems. The risk is not additive where one failure simply adds another. It is multiplicative, where each failure accelerates the next, amplifying impact, compressing response time, and overwhelming recovery assumptions. In such an environment, resilience is determined not by the strength of individual sectors, but by the survivability of the whole system.

Chapter 16 — Exposure: How Many Systems Are at Risk?

EMP risk often feels abstract because it is discussed in institutional terms, the grid, the cloud, critical infrastructure. These phrases sound monolithic, distant, and almost conceptual. In doing so, they obscure the true nature of the exposure.

Modern society does not depend on a handful of systems. It depends on millions of electronic components operating continuously, often without protection and frequently without redundancy. The risk is not concentrated in a single failure point. It is distributed across an enormous, interdependent electronic ecosystem.

Nowhere is this more evident than in information technology and data centers. The United States alone contains an estimated 5,000 to 6,000 data centers, ranging from hyperscale cloud facilities to enterprise and colocation environments. Within each facility reside layers of vulnerable electronics: servers and processors, storage arrays, network switches and routers, power distribution units and uninterruptible power supplies, automatic transfer switches, and environmental monitoring systems. Even a modest data center may contain more than 50,000 electronic components susceptible to electromagnetic disruption. Hyperscale facilities contain orders of magnitude more. Redundancy improves availability, but it does not guarantee survivability.

The energy sector exhibits similar scale. The U.S. electric grid comprises approximately 7,300 power plants, more than 160,000 miles of high-voltage transmission lines, and roughly 55,000 substations. Each relies on protective relays, SCADA systems, communications equipment, and control electronics. Many of the most critical assets, large power transformers, are custom-built, globally sourced, and require replacement timelines measured in twelve to twenty-four months. When these components fail, recovery is constrained not by intent, but by manufacturing capacity.

Communications infrastructure extends this dependency further. The United States operates roughly 400,000 cellular sites, tens of thousands of switching facilities, satellite ground stations, and vast networks of fiber amplification and routing electronics. EMP does not need to topple towers to silence networks. It disables the coordination electronics and power systems that allow communications to function at all.

Financial services illustrate the fragility of trust. Modern financial infrastructure depends on synchronized timing systems, often GPS-based, transaction processing servers, secure networking equipment, and data integrity controls. An electromagnetic disruption does not eliminate money. It eliminates verification. When electronic trust is lost, markets stall regardless of capital availability.

Healthcare, water, and transportation systems compound the exposure. The United States includes approximately 6,100 hospitals and tens of thousands of clinics, each reliant on electronic diagnostics, records, and control systems. Water and wastewater services encompass roughly 50,000 community systems dependent on electronically controlled pumping and treatment infrastructure. Transportation relies on electronic navigation, signaling, and logistics coordination across aviation, rail, and ports.

Across these sectors, the number of EMP-exposed components reaches into the tens, if not hundreds of millions. This exposure is not theoretical. It is embedded in the architecture of modern life.

The risk is not hypothetical. It is structural.

Chapter 17 — Consumer Voice: Why Risk Is Not Abstract

EMP risk is often framed as a national security issue or an abstract enterprise concern, something managed by governments, utilities, or large institutions. That framing misses a critical truth: the consequences of electromagnetic disruption are ultimately borne by individuals and households. Consumers do not experience infrastructure as systems or sectors. They experience it as services and products that are assumed to work. Electricity is expected to be available. Payments are assumed to clear. Fuel, healthcare, communications, and access to money are taken for granted until they are not.

When those services degrade, even partially, the impact is immediate and personal. EMP disruption does not need to destroy infrastructure to alter daily life. Intermittent failure is enough. Electronic payments stop working or work unpredictably. Access to cash becomes unreliable. Healthcare services are delayed or canceled. Fuel availability tightens. Communications fail at the very moment people attempt to seek information or assistance. These effects compound rapidly. What begins as inconvenience becomes stress. Stress turns into disruption. Disruption, sustained over time, becomes dislocation, forcing individuals and families to make decisions they never expected to face.

Many consumers assume that risk is managed upstream, by utilities, governments, cloud providers, or financial institutions. This belief creates a false sense of distance. Households are embedded in the same dependency chains as enterprises and governments. Cloud services authenticate access to personal accounts. Digital identity systems authorize transactions. Financial networks process everyday payments. Logistics platforms determine the availability of food, fuel, and medicine. These systems operate continuously and invisibly, linking consumers to infrastructure sectors in real time. Physical distance offers no insulation from electronic dependency.

Historically, large-scale investments in resilience follow visible failure. Building codes evolved after earthquakes. Flood control expanded after catastrophic inundation. Aviation security transformed after acts of terrorism. EMP has not yet produced a public catalyst dramatic enough to force similar action. But the absence of attention does not imply the absence of risk. It reflects a gap between exposure and awareness, a gap that persists only until disruption makes it undeniable.

Consumers are not powerless in this equation. Public expectation shapes institutional priorities. Awareness drives accountability. Individuals can engage elected representatives, support preparedness-focused policies, ask service providers about continuity planning, and demand transparency around infrastructure dependencies. Preparedness is not solely a government or enterprise responsibility. It is a shared one, because the consequences of failure are shared as well.

Chapter 18 — Why Awareness Precedes Action

Infrastructure resilience is not determined by technology alone. It is shaped by the choices societies make about what to prioritize, what to fund, and what risks are taken seriously enough to confront. Technical capability matters, but it only translates into protection when leadership aligns awareness with action. Prioritization determines which systems are hardened and which are left exposed. Funding decisions reflect what leaders believe is worth protecting. Political will governs whether known vulnerabilities are addressed proactively or deferred until failure makes them unavoidable.

Electromagnetic risk has not yet produced a moment of public clarity. There has been no singular event dramatic enough to force collective recognition, no unmistakable failure that compels immediate, sustained response. As a result, EMP remains understood in fragments, acknowledged in reports, discussed in hearings, and quietly set aside in favor of more visible or immediate concerns. This chapter exists to close that gap deliberately, without fear and without exaggeration, by connecting what is already known to what has not yet been done.

Awareness alone does not guarantee action. History makes that clear. But history is equally clear about the inverse: without awareness, action almost never occurs. Preparation begins not with mandates or funding, but with recognition, a recognition that the risk is real, the exposure is systemic, and the cost of waiting grows silently over time.

Case Study 3 — When Trust Depends on Electronics

During a continuity review of a financial market infrastructure environment, senior leadership examined scenarios involving extended outages of trading and settlement systems. Existing contingency plans emphasized deferred settlement, manual intervention, and regulatory forbearance as mechanisms to absorb disruption. On paper, the institution appeared well positioned. Capital adequacy was strong. Liquidity buffers were sufficient. Confidence in the system's ability to withstand disruption seemed justified.

That confidence rested on a set of assumptions that had become so familiar they were rarely examined directly. Leaders assumed electronic systems could be restored incrementally, allowing partial functionality to return over time. They believed manual processes could temporarily substitute for automated ones, at least long enough to bridge recovery. They expected market participants to tolerate delays as long as capital remained intact, and they assumed trust could be preserved through policy actions, regulatory assurances, and communication. Electromagnetic risk, when it surfaced at all, was treated not as a distinct category of failure, but as an extreme extension of outage scenarios the organization already believed it understood.

A deeper analysis challenged those assumptions. What emerged was a recognition that trust itself was electronically mediated. Transaction verification depended on synchronized clocks. Settlement relied on real-time data integrity and continuous communications. Counterparty confidence was anchored not in policy statements, but in visibility, the ability to confirm, in near real time, that obligations were being met. Manual processes, while theoretically available, lacked the scale, speed, and coordination necessary to support modern market volumes. They were not a substitute for automation; they were a relic of a different era.

Even short-term loss of electronic verification introduced ambiguity that policy alone could not resolve. Markets did not simply slow. They stalled. Participants could not distinguish between temporary disruption and systemic failure. Liquidity hesitated. Risk tolerance collapsed. The absence of reliable confirmation became more destabilizing than the absence of capital itself.

What mattered most, leadership realized, was not the immediate financial loss associated with delayed transactions. It was the erosion of trust. Once confidence degraded, restoring it would take far longer than restoring systems. Technical recovery could be measured in days or weeks. Trust recovery would be measured in months or years.

The lesson was unmistakable. Modern markets run on confidence, and confidence runs on electronics. Protecting the mechanisms of verification and trust is as critical as protecting capital itself. In highly interconnected financial systems, loss of verification halts markets faster than loss of capital ever could.

PART VII — AWARENESS WITHOUT OWNERSHIP

Chapter 19 — Congressional Awareness

EMP risk is not a secret in Washington.

For more than two decades, Congress has examined electromagnetic threats through commissions, hearings, defense panels, and scientific studies. Across administrations and party lines, official reports have consistently warned of vulnerabilities in the electric grid, the fragility of large transformers, exposure of satellite systems, and the potential for cascading failure across critical infrastructure sectors.

The science is settled.
The vulnerability is acknowledged.
What remains unresolved is execution.

Congress first addressed electromagnetic pulse risk in a sustained and formal manner with the creation of the *Commission to Assess the Threat to the United States from Electromagnetic Pulse (EMP) Attack*. The commission was authorized under the *Floyd D. Spence National Defense Authorization Act for Fiscal Year 2001* (Public Law 106-398). Its mandate was notable not only for its scope, but for its recognition that EMP posed a threat not just to military systems, but to civilian infrastructure as well, a distinction that would prove central to every finding that followed.

Recognizing the importance of the issue, Congress reauthorized the commission through the *National Defense Authorization Act for Fiscal Year 2006* (Public Law 109-163), extending its work and reinforcing its relevance. Over the course of its mandate, the commission produced several major reports, including the *2004 Executive Report* and the *2008 Report on Critical National Infrastructures*.

The conclusions were unambiguous. EMP was identified as a high-impact, low-frequency threat. Civilian infrastructure was found to be highly vulnerable. Protective measures were deemed technically feasible and cost-effective. The primary barrier to action was not scientific uncertainty or technical impossibility, but organizational inertia and lack of ownership.

Commission findings were briefed to key congressional bodies, including the House Armed Services Committee, the Senate Armed Services Committee, and the House Homeland Security Committee. Despite this clarity, implementation remained voluntary. Awareness did not translate into mandated protection.

Congress revisited EMP risk periodically through hearings rather than binding legislation. A notable example occurred on May 13, 2014, when the House Committee on Homeland Security's Subcommittee on Cybersecurity, Infrastructure Protection, and Security Technologies convened a hearing titled *"Electromagnetic Pulse (EMP): Threat to Critical Infrastructure."* Testimony reaffirmed earlier conclusions regarding vulnerability and feasibility of protection. Once again, however, no enforceable mitigation requirements followed.

A more significant milestone arrived with *Executive Order 13865*, issued on March 26, 2019, titled *"Coordinating National Resilience to Electromagnetic Pulses."* The order directed federal agencies to identify EMP risks, assess infrastructure resilience, coordinate preparedness efforts, and engage public-private partnerships. Congress later codified elements of this directive in the *National Defense Authorization Act for Fiscal Year 2020* (Public Law 116-92).

Yet even this step stopped short of imposing enforceable requirements on privately owned infrastructure, the majority of systems upon which national continuity depends.

Viewed collectively, a pattern emerges. In 2001, the EMP Commission was established. In 2004 and 2008, authoritative findings were released. In 2006, the commission was reauthorized. In 2014, congressional hearings reaffirmed known risks. In 2016, GAO oversight reports echoed similar concerns. In 2019, executive action acknowledged the threat. In 2020, those provisions were partially codified. And today, studies continue while implementation remains limited.

Recognition without compulsion.

Congress has done much. It has commissioned rigorous studies, convened hearings, acknowledged EMP risk, and incorporated electromagnetic resilience into national policy frameworks. What it has not done is mandate EMP protection standards for private infrastructure, fund large-scale civilian hardening, or resolve the fundamental question of ownership and accountability. This is not a failure of awareness. It is a failure of sustained ownership.

Chapter 20 — Why Reports Don't Equal Protection

Reports are necessary. They are not sufficient.

For decades, electromagnetic risk has been studied, documented, and debated. Reports have clarified the physics. White papers have cataloged vulnerabilities. Hearings have acknowledged exposure. Together, they have built a body of knowledge that leaves little ambiguity about the nature of the threat.

But none of these activities harden infrastructure.

No report has ever strengthened a transformer.
No white paper has shielded a data center.
No hearing has installed grounding, protected control electronics, or validated that systems will survive electromagnetic stress.

EMP mitigation does not occur in documents. It occurs in physical systems.

Real resilience requires engineering, a careful design that acknowledges electromagnetic realities rather than assuming them away. It requires capital investment, often upfront and often without immediate visible return. It requires validation and testing, because protection that has not been tested is indistinguishable from protection that does not exist. And it requires sustainment, because systems evolve, configurations change, and protection degrades over time if it is not maintained.

These activities are inherently difficult to justify in environments where success is defined by absence. When nothing happens, preparation is invisible. When systems continue to function, resilience appears unnecessary. This creates a structural bias against action.

Political and organizational systems tend to reward visibility, immediacy, and measurable short-term outcomes. EMP resilience offers none of these. Its value is revealed only when failure is avoided, often quietly, and often without public recognition. As a result, electromagnetic risk is frequently acknowledged in principle while postponed in practice.

This dynamic produces a dangerous condition: familiarity without urgency.

Organizations become accustomed to hearing about EMP risk without acting on it. Over time, deferral becomes normalized. The threat is neither denied nor addressed. It is simply absorbed into the background noise of competing priorities. The danger is not ignorance. The danger is comfort.

Familiarity dulls the impulse to act. Urgency fades not because risk has diminished, but because it has become abstract. And when urgency is absent, execution never follows.

EMP resilience fail because leaders are uninformed, underinformed or are taking a pass. Often, it fails because the systems that govern decision-making reward delay more reliably than preparation. This gap between what is known and what is done is where vulnerability persists.

Until that gap is closed, awareness alone will continue to coexist with exposure.

Chapter 21 — The Limits of Regulation

Even with political will, regulation alone cannot resolve electromagnetic risk.

The majority of U.S. critical infrastructure is privately owned and operated. Regulation, by design, establishes minimum standards intended to ensure baseline safety and compliance, not optimal resilience against low-probability, high-impact threats. It defines what is acceptable, not what is sufficient to survive extreme disruption.

Regulatory frameworks also lag reality. They evolve slowly, often in response to past events rather than emerging threats. Electromagnetic risk, by contrast, advances alongside technology. As systems become more digitized, centralized, and electronically dependent, exposure increases faster than regulatory mechanisms can adapt.

In practice, regulation encourages compliance over engineering. Organizations focus on meeting documented requirements rather than interrogating physical survivability. A system may be fully compliant and still be fundamentally fragile. Checklists are satisfied. Vulnerabilities remain.

Most critically, regulation cannot enforce survivability. It can mandate reporting, audits, and procedural controls, but it cannot guarantee that infrastructure will function under electromagnetic stress. Survivability requires design decisions, capital investment, validation, and sustained ownership, none of which can be fully compelled through rulemaking alone.

Waiting for mandates ensures delay. It shifts responsibility outward and creates a false sense of security, reinforcing the belief that if protection were necessary, it would already be required. History suggests otherwise. Many resilience measures, from building codes to aviation safety, were adopted only after failure exposed their absence.

Electromagnetic resilience therefore must be embraced as a leadership responsibility. It is grounded not in compliance, but in fiduciary duty and stewardship, the obligation to preserve continuity, protect stakeholders, and safeguard long-term enterprise value even when risks are difficult to quantify.

Government involvement is essential. It can convene, coordinate, incentivize, and inform. But it cannot substitute for ownership.

Ultimately, resilience is not granted by regulation.
It is chosen by leadership.

Chapter 22 — The Wake-Up Event

Societies rarely prepare for catastrophic risk in advance. History shows that meaningful action most often follows visible harm, not credible warning. Electromagnetic risk is no exception. Despite decades of study and repeated technical validation, preparedness remains limited because the event that forces recognition has not yet occurred, or, more precisely, has not yet been recognized for what it is.

When that moment comes, it is unlikely to arrive with spectacle. There will be no singular flash, no shared moment of clarity, no immediate declaration that an electromagnetic event has occurred. Instead, the wake-up event will almost certainly appear fragmented, regional, and ambiguous, indistinguishable at first from the kinds of disruptions modern societies have learned to absorb and rationalize.

It may begin as a prolonged power outage that exceeds restoration estimates. It may surface as unexplained infrastructure failures that do not align with known fault patterns. It may coincide with a severe solar storm that initially appears manageable. It may take the form of a regional communications collapse that disrupts emergency coordination and commercial activity without an obvious cause. Each of these scenarios, viewed in isolation, is familiar. None is sufficient, on its own, to trigger immediate recognition.

What will distinguish the wake-up event is not the initial failure, but the failure to recover.

As hours become days and days stretch into weeks, confidence begins to erode. Restoration timelines slip. Backup systems that were assumed to provide margin prove insufficient or unavailable. Communications remain degraded. Data integrity is questioned. Mutual aid agreements strain under simultaneous demand. Investigations multiply, but answers remain elusive. Leaders are forced to explain, not why a failure occurred, but why redundancy did not deliver the resilience it promised.

Only then does recognition begin to emerge.

The Silent Flash

The realization will not arrive as a declaration, but as a pattern: failures that cut across sectors, recovery efforts that stall despite resources, and dependencies that were assumed external but remain unavailable far longer than expected. What initially appeared as technical malfunction reveals itself as systemic fragility.

Electromagnetic risk does not announce itself through destruction. It reveals itself through absence of power, of communication, of coordination, and ultimately of trust. The wake-up event is not defined by what breaks first, but by what does not come back.

By the time that distinction is understood, the window for proactive protection will have already closed.

Part VIII – FAILURE MADE REAL

When societies evaluate risk, they often rely on historical financial benchmarks: the costliest natural disasters, industrial accidents, and wars. Major hurricanes, earthquakes, floods, and wildfires typically generate economic losses in the $10 to $200 billion range per event, even when human impact is severe.[1] These events are destructive, but their effects are generally geographically bounded, and the systems required for recovery, power grids, financial markets, logistics networks, and governance, remain largely functional outside the affected region.

Large-scale wars introduce higher aggregate costs, often reaching $1 to $5 trillion over time.[2] However, these losses typically accrue over years and occur alongside functioning industrial capacity, operational financial systems, and intact global trade networks that enable sustained recovery and economic adaptation.

Electromagnetic disruption alters this model fundamentally. Severe EMP scenarios are not defined primarily by visible destruction, but by the simultaneous degradation of the systems required for response, recovery, and economic continuity. Credible government and industry assessments estimate that a large-scale electromagnetic event could produce multi-trillion-dollar impacts, ranging from approximately $2 trillion to more than $10 trillion, depending on scope, duration, and restoration constraints.[3-5] These figures reflect not only physical damage to infrastructure, but prolonged loss of economic output, disruption of financial settlement and payment systems, impairment of healthcare and public services, and cascading failures across multiple critical infrastructure sectors.

Unlike conventional disasters, EMP does not impose a single, discrete cost. It degrades the mechanisms that normally limit losses and accelerate recovery. Power, communications, transportation, manufacturing, and financial verification systems may be impaired simultaneously, extending disruption from weeks into months or years.[6] In economic terms, EMP represents not merely another category of disaster, but a systemic risk amplifier, one capable of transforming disruption into sustained national and global economic instability.

The scenarios and economic impact ranges presented in this chapter are hypothetical planning constructs. They are not predictions of specific events, timelines, or outcomes. Quantitative values are expressed as orders of magnitude to reflect uncertainty and variability across geography, preparedness, system architecture, and response conditions. Actual outcomes would differ in any real event.

The purpose of these scenarios is not to forecast the future, but to support informed, risk-aware decision-making and resilience planning.

Scenario 1 — The Parking Lot Event

Localized EMP, Disproportionate Damage

The event began quietly. A truck-mounted, non-nuclear electromagnetic device was activated near a dense telecommunications corridor serving a major metropolitan area. There was no explosion, no visible damage, and no immediate alarm. The physical footprint was small. The consequences were not.

Within minutes, routing instability appeared inside a carrier switching facility. Automated failover mechanisms engaged as designed, redistributing traffic away from affected equipment. But those mechanisms relied on shared upstream infrastructure, and the rebalancing itself propagated instability beyond the immediate area. What initially appeared to be a localized anomaly spread across the metro network.

Enterprise authentication systems began to fail intermittently. Applications remained technically available, yet users could not log in reliably. Sessions dropped. Transactions timed out. Systems were neither clearly up nor clearly down. The ambiguity proved more damaging than a clean outage.

For the telecommunications provider, service-level agreement penalties and customer credits accumulated rapidly, reaching an estimated **$5 to 15 million**. More damaging than the immediate loss was the erosion of credibility. A facility marketed as "carrier-grade" had failed in a way customers could not reconcile. Over the following year, customer attrition increased by **5 to 10 percent**, while future contract opportunities exceeding **$50 million** quietly evaporated.

Enterprise customers experienced direct revenue losses as transactions stalled and billing was delayed. Mid-sized firms absorbed **$1 to 5 million** in losses, alongside thousands of idle staff-hours and six-figure remediation costs. Customer churn rose **3 to 7 percent**, and boards ordered resilience reviews that delayed growth initiatives.

Healthcare outpatient systems were forced to cancel procedures as scheduling and diagnostic systems failed. Losses of **$2 to 4 million** followed, alongside regulatory reporting and lingering declines in patient confidence.

Municipal traffic systems failed intermittently, producing congestion and emergency response delays. Regional productivity losses exceeded **$10 million**, and public trust in digital infrastructure eroded.

The device itself cost less than **$100,000**. The damage reached into the **tens of millions**.

Closing insight:
The most damaging effect was not the outage itself, but the loss of confidence. Localized EMP does not stay local when systems are tightly coupled and auto-rebalance without independence.

Scenario 2 — The Substation Control Failure

Power Without Control

A small unmanned aerial system delivered a localized electromagnetic disruption near a regional electrical substation. The event did not destroy physical infrastructure. Instead, it disrupted control electronics.

Protective relays began behaving unpredictably. Operators lost reliable telemetry and could not confirm safe switching states. Power was not uniformly lost, it became unstable. Partial restoration attempts triggered downstream equipment damage, extending outages.

As instability persisted, dependent systems began to fail. Water pumping stations lost reliable control, prompting boil advisories. Industrial refrigeration units shut down. Healthcare clinics closed as power quality fluctuated beyond tolerances.

The energy provider incurred **$20 to 40 million** in emergency response costs, equipment damage, and overtime. Regulatory scrutiny intensified, and capital previously earmarked for modernization was redirected toward remediation.

Water utilities absorbed **$3 to 6 million** in emergency sourcing and billing losses, alongside long-term damage to public trust. Industrial customers lost **$10 to 30 million** in spoiled inventory and experienced **10 to 15 percent** contract churn in the following year.

Healthcare clinics lost **$1–3 million** in revenue over several days, while nearby hospitals absorbed surges that strained staff and systems.

The outage ended, but the economic impact persisted.

Closing insight:
Electricity alone is not resilience. Control, visibility, and confidence determine whether power restores stability, or compounds risk.

Scenario 3 — The Data Center Edge Collapse

When Security Becomes the Failure Point

A directed-energy electromagnetic event affected a multi-tenant colocation campus. Core compute infrastructure remained largely intact. Edge routing and security systems did not.

Firewalls, identity brokers, and inspection appliances failed asymmetrically. Some traffic passed. Some failed silently. Authentication tokens validated intermittently. Data integrity could not be verified. Several tenants chose to shut systems down, not because they were offline, but because they could not prove correctness.

Failover regions existed, but access depended on identity and security paths that were also impaired. Backup data was available, yet restoration required systems that were unreachable.

The colocation provider issued customer credits and absorbed churn totaling **$15 to 30 million**, alongside a **10 to 20 percent** reduction in new customer growth as confidence eroded.

Financial services tenants missed settlements, halted trading, and absorbed **$25 to 100 million** in direct losses per institution. Liquidity stress triggered emergency capital actions and regulatory scrutiny.

Healthcare clearinghouses accumulated claims backlogs exceeding **$500 million** in deferred reimbursement, damaging provider relationships. Logistics platforms experienced **8 to 12 percent** customer churn as delivery windows were missed and inventory sat idle.

Closing insight:
Security controls that cannot survive disruption become single points of failure. Trust in correctness, once lost, halts systems faster than physical damage.

Scenario 4 — The Regional EMP Event

Progressive Economic Failure

A regional electromagnetic event disrupted electronics across multiple counties. Power systems degraded unevenly as transformers and control electronics failed. Telecommunications networks followed as backup power was exhausted.

Recovery bottlenecks compounded. Fuel deliveries prioritized hospitals, delaying commercial restoration. Replacement electronics were globally backordered. Staff absenteeism increased as personal hardship mounted.

Small and mid-sized businesses lost **30 to 60 percent** of annual revenue. Between **20 and 40 percent** closed permanently. Energy and telecom providers incurred **hundreds of millions of dollars** in repair costs and deferred modernization for years.

Healthcare systems canceled procedures totaling **$50 to 150 million** and operated under crisis standards of care. Workforce attrition accelerated. Consumers lost wages and access to essentials, driving population displacement.

The region recovered unevenly. Economic scars remained visible years later.

Closing insight:
Regional EMP does not collapse society. It reshapes economies, redistributes population, and permanently alters competitive landscapes.

Scenario 5 — The High-Altitude EMP Event (HEMP)

Survivable, but Transformative

A high-altitude nuclear detonation produced a widespread electromagnetic pulse. Electronics failed across sectors simultaneously. Power, communications, finance, and logistics degraded together.

Systems were not uniformly destroyed. Many remained physically intact but unusable. Coordination failed faster than infrastructure.

Enterprises lost **10 to 30 percent** of annual revenue, missed market cycles, and suffered permanent competitive disadvantage. Survivability, not size, determined recovery order.

Cloud providers experienced **5 to 10 percent** large-customer churn as assumptions of "always-on" availability collapsed. Financial markets required policy intervention to stabilize liquidity measured in the **hundreds of billions.**

Consumers experienced prolonged income disruption and shifted behavior away from digital-only dependency. Governments absorbed **trillions** in stabilization costs as infrastructure governance was reshaped.

Society endured. Momentum did not.

Closing insight:
National-scale EMP does not end civilization. It redistributes wealth, trust, and power, and defines which institutions emerge intact.

The Uncomfortable Truth These Scenarios Reveal

Across every scale, the pattern is consistent:

- losses exceed repair costs

- reputational damage outlasts outages

- trust recovers slower than systems

- some organizations never return

EMP does not need to massively destroy infrastructure to destroy value. It only needs to remove coordination long enough for confidence to fail.

The most dangerous outcome is not collapse.

It is **being wrong about how bad it would be**.

Chapter 24: Impact to Investment: ROI of Electromagnetic Resilience

Organizations rarely fail because leaders ignored risk entirely. They fail because the **economic framing of that risk was incomplete.** EMP resilience has historically been discussed in terms of improbability or national security. That framing obscures the reality decision-makers care about:

What is the financial consequence of being wrong?

The scenarios in the previous chapter demonstrate a consistent pattern: electromagnetic disruption does not require catastrophic destruction to produce **outsized economic loss.** Value is lost through downtime, erosion of trust, regulatory exposure, customer attrition, and missed opportunity, long before infrastructure is physically rebuilt.

The correct question is not *"How likely is an EMP event?"* It is *"What happens to us if one occurs — and we are unprepared?"*

The Asymmetry of Cost vs. Consequence

Across industries, a striking asymmetry appears:

Category	Order of Magnitude
Cost to implement targeted EMP resilience	0.5–3% of annual IT or infrastructure spend
Loss from a localized EMP disruption	5–20% of annual revenue
Loss from a regional EMP disruption	30–60% of annual revenue
Probability of full recovery without preparation	Uncertain

Category	Order of Magnitude
Probability of permanent damage with no preparation	High

The economic logic is straightforward:

a small, bounded investment mitigates an unbounded downside.

This is not speculative risk transfer. It is margin preservation.

ROI by Scenario Class: Why Resilience Pays Before It Is Needed

Electromagnetic resilience investments are often evaluated using the wrong lens. They are compared against routine IT upgrades or incremental cybersecurity controls, rather than against the magnitude and duration of losses that follow systemic disruption. When framed correctly, the return on resilience is not subtle. It is decisive.

Localized EMP - Facility or Node-Level Disruption

At the facility level, resilience investments are both achievable and disproportionately effective. A mid-sized enterprise can materially reduce exposure through hardened power conditioning and grounding, shielding of critical electronics, protected control and monitoring systems, and disciplined validation and testing. These measures do not attempt to harden everything. They focus on preserving coordination, visibility, and control.

For many organizations, the capital required to implement this level of protection falls between **$250,000 and $2 million**, often comparable to a single major software deployment or a few weeks of operating expense.

The losses avoided, however, are orders of magnitude larger. Localized EMP events could produce **$5 to 25 million** in direct downtime losses, driven by halted operations, failed transactions, and emergency remediation. Customer churn of possibly **3 to 10 percent** follows as trust erodes, and reputational damage lingers for years, quietly suppressing growth and future contract opportunities.

The logic is straightforward: a one-time investment equivalent to weeks of "other priority" expense prevents losses equivalent to years of profit. This is not a bet on probability. It is a hedge against inevitability in an increasingly dense and exposed environment.

Regional EMP — Multi-Sector Disruption

At regional scale, the role of resilience shifts from loss prevention to survival advantage. Organizations cannot insulate themselves from every external dependency, but they can preserve enough internal capability to continue operating while others fail.

Effective regional resilience focuses on protecting priority systems rather than attempting comprehensive hardening. Independent power conditioning, survivable communications paths, and manual-operable fallback for critical functions create operational margin when upstream services degrade or disappear.

For large enterprises or critical facilities, these measures typically require **$2 to 10 million** in investment. That figure often appears large, until contrasted with the consequences of inaction.

Regional EMP disruption can drive **$50 to 500 million** in revenue loss for exposed organizations. Small and mid-sized enterprises face existential risk, with closure rates rising sharply after prolonged outages. Even large firms experience long-term market displacement as customers migrate to competitors who remain functional.

Here, the return on investment is not measured by avoided downtime alone. It is measured by preserved optionality, the ability to operate, serve customers, and make decisions while competitors are immobilized. In disrupted markets, continuity becomes a competitive weapon.

National-Scale EMP (HEMP)

At national scale, no organization escapes unscathed. The question is not whether loss occurs, but who recovers first and who never fully does.

Resilience investments at this level are necessarily focused. They prioritize survivability of foundational systems: identity, control, coordination, and data integrity. They include protected command functions and pre-positioned recovery capability that allows operations to resume before normal conditions return.

These investments typically represent a **single-digit percentage of annual infrastructure budgets**, a level often absorbed without material impact on growth initiatives.

The losses they mitigate are existential. National-scale EMP events can drive **10 to 30 percent annual revenue loss**, permanently alter competitive positioning, and inflict reputational harm that cannot be repaired through marketing or policy alone. Organizations that fail to preserve operational trust do not simply lose a year, they lose relevance.

In this context, resilience does not eliminate loss. It determines survivability.

Why Traditional ROI Models Fail

Conventional ROI frameworks are poorly suited to electromagnetic risk because they rely on assumptions that do not hold under EMP conditions. They assume failures are predictable, downtime is bounded, external services remain available, and trust is recoverable once systems restart.

EMP violates all four.

Electromagnetic events create correlated failure across systems, extend downtime due to global supply constraints, collapse dependencies on power, communications, and identity, and erode trust in ways that outlast technical recovery. Losses do not stop when systems come back online.

As a result, resilience ROI cannot be evaluated as a cost-saving exercise. It must be evaluated as **loss avoidance and survival probability**.

The organizations that invest are not betting on fear. They are buying time, credibility, and the ability to act when others cannot. That is not speculative value. It is strategic insurance against a class of failure that does not forgive unpreparedness.

Resilience as Fiduciary Duty

Boards and executives are legally and ethically obligated to:

- understand material risk

- act on known vulnerabilities

- protect long-term enterprise value

EMP risk meets every criterion of materiality:

- the physics are known

- the vulnerabilities are documented

- the consequences are severe

- mitigation is technically feasible

Choosing not to act is not neutral.
It is a decision to **accept asymmetric downside**.

The Real ROI: Time, Trust, and Control

In every scenario examined, organizations that fared best shared three traits:

1. **They bought time** - systems stayed functional longer.

2. **They preserved trust** - data and operations remained verifiable.

3. **They retained control** - decisions were made, not guessed.

Those outcomes cannot be purchased during a crisis.
They must be engineered beforehand.

The Bottom Line

EMP resilience is not an insurance policy. It is not about preventing all loss. It is about preventing **irreversible loss**. When evaluated honestly, the return on investment is not marginal. It is decisive. The most expensive resilience investment is the one made after it is too late.

PART IX - THE RESPONSIBILITY WE ACCEPT

Chapter 25 — Why It Won't Be "The Big One"

There is a persistent belief that meaningful action only follows unmistakable catastrophe. History suggests otherwise. Large, singular disasters are rare, while smaller failures are common, and often far more instructive. They expose fragility without overwhelming response capacity, forcing uncomfortable questions while still leaving room to act. The most dangerous assumption is not that risk will emerge without warning, but that the first warning will be obvious.

In practice, early indicators are subtle and easily dismissed: a delayed market settlement, a regional grid instability, a communications outage that lasts longer than expected. Each incident can be rationalized in isolation. Together, they form a pattern. The greatest risk is not that an electromagnetic event will occur without warning. It is that warning will be misinterpreted, normalized, or ignored until response options narrow or disappear entirely.

Chapter 26 — The Question We'll Ask After

After disruption subsides, after hearings, inquiries, and reports, a single question will dominate public and institutional discourse: *Why did we know, and fail to act?* EMP risk is not hidden. The physics are understood. The vulnerabilities are documented. The consequences are foreseeable. What remains unresolved is accountability.

Explanations will cite budgets, mandates, and competing priorities. Many of these explanations will be accurate. None of them will be sufficient. Leadership is not measured by awareness; it is measured by preparation.

The United States was built through determination, ingenuity, and an unwavering commitment to endure. We have faced adversity before, war, disaster, and profound loss, and emerged stronger because we chose preparation over complacency. Our national strength is not accidental. It is the result of deliberate choices to protect what matters.

Today, electromagnetic risk challenges us in a different way. It threatens the infrastructure that sustains our economy, our security, and our daily lives, often invisibly and without warning. This risk does not respect political boundaries or competing interests. It affects every American. The choice before us is stark: will we address this vulnerability with foresight and unity, or allow distraction, fragmentation, and short-term priorities to leave our nation exposed? Will we invest proactively in protecting the systems that sustain modern life, or wait for failure to force action upon us?

Resilience is not partisan. It is foundational. Protecting the continuity of our nation is not optional, it is a responsibility. History will not judge us by what we debated, but by whether we acted before the cost of inaction became irreversible. This book exists to make that question unavoidable, before events force it upon us.

Chapter 27 — Who Must Be Protected

Electromagnetic resilience is not about protecting everything. It begins with prioritization. Certain systems are foundational: power enables all others, communications coordinate response, financial systems preserve trust, healthcare preserves life, and water sustains populations. These systems form the minimum conditions for continuity.

Within organizations, the same principle applies. Not every system is critical. Some failures are tolerable. Others are existential. Avoiding these decisions does not eliminate risk; it transfers it, often to those least able to absorb the consequences. Leadership is defined not by whether prioritization occurs, but by whether it is deliberate or deferred.

Chapter 28 — How Resilience Actually Works

Resilience is not a slogan. It is a discipline. Effective electromagnetic resilience follows a lifecycle: assessment to identify what truly matters; design to engineer survivability rather than availability; implementation to translate theory into reality; validation to confirm protection works; and sustainment to ensure it remains effective over time. Skipping steps produces symbolic protection rather than real capability.

Resilience is quiet work. Its success is measured not by visibility, but by systems that continue to function when others fail.

It is tempting to view electromagnetic resilience as someone else's responsibility - energy providers, utilities, or governments. This view is understandable, but incomplete. Infrastructure does not fail all at once, and it does not recover all at once. Outcomes are shaped by how long downstream organizations can remain functional while upstream services are degraded. That margin is something organizations can influence today.

Remediation does not mean solving national grid fragility. It means protecting what falls within direct control: internal power conditioning, critical electronics and control systems, data survivability, and coordination capability under degraded conditions. These measures do not make organizations independent, but they do make them less brittle. Resilience is cumulative.

Waiting is not a neutral act. Waiting assumes that upstream protection will arrive soon, that it will be comprehensive, and that it will fail gracefully. History suggests otherwise. Infrastructure improvements take decades, while organizations continue to digitize and centralize, often increasing exposure faster than resilience improves. Waiting is a decision to allow dependency to grow unchecked.

Acting now is rational. Resilience is not about predicting catastrophe; it is about preserving time, decision space, and optionality. Organizations that endure are not those that guessed timing correctly, but those that preserved margin. Internal resilience is not a substitute for national resilience. It is a bridge to it.

Chapter 29 — Stewardship in an Age of Accelerated Dependency

Every generation inherits risks it did not create. Electromagnetic pulse is not new; what is new is how profoundly modern society depends on the systems it threatens. Contemporary leadership rightly prioritizes growth, efficiency, and innovation, yet these priorities alone are insufficient in an age defined by fragility. Leadership today must also embody stewardship, the willingness to accept responsibility for risks that may never materialize within one's tenure but would be catastrophic if they did. EMP exposes a fundamental imbalance: the cost of preparation is immediate, visible, and often inconvenient, while the cost of inaction is deferred, diffuse, and irreversible. True leadership is defined by the ability to accept this imbalance and act anyway.

One of the most notable expansions in dependency on infrastructure is the emergence of Artificial Intelligence (AI). AI is not an incremental evolution of technology; it is a force multiplier for centralization, automation, and continuous operation. Modern AI systems depend on dense computational environments, uninterrupted power, synchronized data, and persistent connectivity. As these systems proliferate, manual fallback diminishes, human improvisation atrophies, and operational models increasingly assume uninterrupted continuity.

In this context, electromagnetic disruption does not merely disable machines, it removes coordinating intelligence. When electronic systems fail, AI-driven decision-making, automation, and optimization fail with them. As AI becomes embedded across finance, healthcare, logistics, and critical infrastructure, the consequences of electronic failure scale exponentially. Efficiency improves, but resilience does not follow automatically.

AI can, however, serve as a powerful tool for resilience, if it is chosen deliberately. When applied with intent, AI can help model interdependencies, expose hidden failure modes, optimize protection strategies, and support decision-making under degraded conditions.

Without architectural diversity, protected infrastructure, and human-in-the-loop governance, however, AI accelerates fragility rather than mitigating it.

Technology reflects values. If efficiency alone is prioritized, fragility advances at machine speed.

Conclusion — The Silent Flash to Stewardship

EMP does not arrive with spectacle. There is no siren, no shockwave, no shared moment of recognition. Instead, there is silence. Screens do not light. Systems do not respond. Connections fail to reconnect, not with chaos, but with absence.

That silence reveals a truth that is easy to forget: resilience is not built during crisis. It is built quietly and deliberately, long before it is ever tested. When systems continue to function, preparation is invisible. When they do not, its absence becomes undeniable.

History rarely remembers the disasters that did not occur. It remembers the cost of neglect, the moments when warnings were known, vulnerabilities documented, and action deferred. EMP may never occur, and that should be welcomed. But the absence of disaster does not justify inaction; it merely postpones judgment.

The question that remains is simple, and it is unavoidable. When faced with a known catastrophic risk, will we choose foresight, or convenience?

The silent flash does not care how we answer.
History will.

NOTES (ENDNOTES)

Chapter 1 — The Accident That Changed Everything

1. U.S. Department of Defense, Starfish Prime nuclear test reports and related declassified materials (1962–1963).

2. Ceruzzi, Paul E., A History of Modern Computing (MIT Press).

3. National Research Council, Severe Space Weather Events—Understanding Societal and Economic Impacts (National Academies Press, 2008).

4. Abbate, Janet, Inventing the Internet (MIT Press).

Chapter 3 — The Two Sources: Nature and Intent

1. National Research Council, Severe Space Weather Events—Understanding Societal and Economic Impacts (National Academies Press, 2008).

2. General historical account of the 1859 Carrington Event and telegraph impacts (verification and preferred primary source selection recommended for final publication).

Chapter 10 — Nation-States and Strategic Doctrine

1. Jacob Stokes, *China's High-Altitude Electromagnetic Pulse Weapons, Cyber Warfare, and Nuclear Deterrence,* Johns Hopkins School of Advanced International Studies, https://jsis.washington.edu/news/chinas-high-altitude-electromagnetic-pulse-weapons-cyberwarfare-and-nuclear-deterrence/

2. *New Physical Principles Weapons,* Wikipedia (summarizing Russian doctrinal concepts and Western analysis), https://en.wikipedia.org/wiki/New_physical_principles_weapons

3. Peter Vincent Pry, testimony before the U.S. House Committee on Homeland Security, *EMP Threat to the United States,* 2017, https://docs.house.gov/meetings/HM/HM09/20171012/106467/HHRG-115-HM09-Wstate-PryP-20171012.pdf

4. James Jay Carafano, *The EMP Threat: Iran and Ballistic Missile Capability,* The Heritage Foundation, https://www.heritage.org/missile-defense/commentary/new-york-times-turns-blind-eye-emp-threat

5. Jean-Dominique Merchet, *Saudi Arabia Would Develop Electromagnetic Pulse Weapons,* Meta-Defense, https://meta-defense.fr/en/2019/01/22/Saudi-Arabia-would-develop-electromagnetic-pulse-weapons-with-the-help-of-Ukrainian-industry/

6.   *Counter-electronics High Power Microwave Advanced Missile Project (CHAMP)*, U.S. Air Force / Boeing overview, https://en.wikipedia.org/wiki/Counter-electronics_High_Power_Microwave_Advanced_Missile_Project

Chapter 12 — Solar Storms: The Threat Without Intent

1. National Research Council, Severe Space Weather Events—Understanding Societal and Economic Impacts (National Academies Press, 2008).

2. NOAA Space Weather Prediction Center (SWPC), educational resources on coronal mass ejections and geomagnetic storms (verification recommended for exact page selection in final publication).

Chapter 15 — Critical Infrastructure and Systemic Exposure

1. U.S. Department of Homeland Security / CISA documentation defining the 16 Critical Infrastructure Sectors (verification recommended for the precise citation page in final publication).

Chapter 16 — Scale of Exposure: How Many Systems Are at Risk?

1. Quantitative infrastructure counts (data centers, plants, miles of transmission, substations, cell sites, hospitals, community water systems) are retained as written from the supplied draft; verification against authoritative sources (DOE/EIA, FCC/NTIA, AHA, EPA) recommended prior to publication.

Chapter 19 — Congressional Awareness

1. Floyd D. Spence National Defense Authorization Act for Fiscal Year 2001, Pub. L. 106-398 (authorization of the EMP Commission).

2. National Defense Authorization Act for Fiscal Year 2006, Pub. L. 109-163 (reauthorization of the EMP Commission).

3. Commission to Assess the Threat to the United States from Electromagnetic Pulse (EMP) Attack, Executive Report (2004).

4. Commission to Assess the Threat to the United States from Electromagnetic Pulse (EMP) Attack, Report on Critical National Infrastructures (2008).

5. U.S. House Committee on Homeland Security, Subcommittee on Cybersecurity, Infrastructure Protection, and Security Technologies, hearing: "Electromagnetic Pulse (EMP): Threat to Critical Infrastructure" (May 13, 2014).

6. Executive Order 13865, "Coordinating National Resilience to Electromagnetic Pulses" (March 26, 2019).

7. National Defense Authorization Act for Fiscal Year 2020, Pub. L. 116-92 (codifying elements related to EMP resilience coordination).

8. U.S. Government Accountability Office (GAO) oversight reporting on electromagnetic threats and/or grid resilience (verification recommended to select the specific GAO report(s) cited).

Chapter 23

1.   **NOAA, FEMA, Munich Re, Swiss Re** — U.S. and global catastrophe loss databases consistently show major natural disasters producing losses in the tens to low hundreds of billions of dollars per event.

- NOAA Billion-Dollar Weather and Climate Disasters

- Swiss Re Sigma Reports

2.   **Congressional Budget Office (CBO); World Bank; SIPRI** — Estimates of major war costs, including World War II and post-9/11 conflicts, place total economic impact in the low-trillion-dollar range over extended periods.

- CBO: *The Costs of Major U.S. Wars*

- SIPRI Military Expenditure Database

3.   **Commission to Assess the Threat to the United States from Electromagnetic Pulse (EMP Commission)** — Multiple reports estimate that long-duration grid failure could result in economic losses measured in trillions of dollars, with recovery timelines extending beyond a year.

- EMP Commission Report (2008, 2017)

4.   **Department of Homeland Security (DHS) / FEMA** — National Risk Management and infrastructure resilience assessments identify EMP and geomagnetic disturbance as high-consequence, low-frequency risks with nationwide economic impact.

- DHS National Infrastructure Protection Plan

- FEMA National Risk Index

5.   **Lloyd's of London / Cambridge Centre for Risk Studies** - Scenario modeling of space weather and electromagnetic disruption estimates global economic losses ranging from **$2–$10+ trillion**, depending on event severity and recovery assumptions.

- Lloyd's: *Solar Storm Risk to the North American Electric Grid*

6.   **North American Electric Reliability Corporation (NERC); EPRI** - Studies highlight transformer damage, manufacturing bottlenecks, and recovery dependencies that extend outage duration far beyond conventional disaster timelines.

- NERC GMD Task Force Reports

- Electric Power Research Institute (EPRI) Grid Resilience Studies

Scenarios and ROI

- National Research Council, *Severe Space Weather Events, Understanding Societal and Economic Impacts*
- Commission to Assess the Threat to the United States from EMP Attack, *Critical National Infrastructures*
- U.S. Department of Energy, *Large Power Transformer Study*
- Lloyd's of London, *Business Blackout* scenario analysis
- FEMA, *National Disaster Recovery Framework*
- GAO reports on grid resilience and continuity
- IEEE publications on EMP and HPM effects
- DHS Cybersecurity & Infrastructure Security Agency (CISA) resilience frameworks

Appendix A — Scenario Modeling and ROI Methodology

Purpose of the Scenarios

The scenarios presented in this book are intended to support risk planning, not prediction. Electromagnetic pulse (EMP) risk is characterized by:

- low observable frequency
- high systemic impact
- limited public post-event data
- strong coupling across infrastructure sectors

Under such conditions, scenario analysis is the appropriate analytical tool. This approach is standard practice in finance, insurance, defense planning, and infrastructure resilience.

Analytical Foundations

The scenarios draw on five established bodies of knowledge:

1. EMP Physics and Engineering Effects

- High-altitude nuclear EMP (HEMP)
- Non-nuclear EMP and high-power microwave (HPM)
- Geomagnetic disturbance (GMD) and solar storm impacts
- Coupling mechanisms into electronics, power systems, and communications

These effects are well documented in government, academic, and military literature and form the technical basis for plausibility.

2. Infrastructure Dependency Mapping

Modern infrastructure is tightly coupled across sectors. Scenario construction explicitly considers:

- power → communications → finance → healthcare → logistics interdependence
- identity, timing, and network dependencies
- cloud concentration and shared physical infrastructure
- loss of coordination versus loss of assets

Dependency mapping is a core method used by DHS, CISA, and critical infrastructure operators.

3. Historical Outage Economics (Non-EMP Analogs)

Because large-scale EMP events lack public civilian datasets, analogous outages are used to bound economic impact, including:

- major grid failures
- regional natural disasters
- prolonged telecommunications outages
- financial market interruptions
- cloud service disruptions

These analogs provide empirical grounding for:

- revenue loss
- customer churn
- business failure rates
- recovery duration
- reputational damage

4. Business Continuity and Disaster Recovery Modeling

The analysis incorporates established findings from:

- FEMA disaster recovery assessments
- insurance catastrophe modeling
- GAO continuity evaluations
- enterprise risk management (ERM) stress testing

This supports assumptions about:

- recovery constraints
- resource scarcity
- coordination breakdown
- workforce availability
- secondary economic effects

5. ROI and Loss-Avoidance Framing

Return on investment (ROI) is evaluated using loss-avoidance logic, not profit generation. This approach is standard for:

- safety engineering
- cybersecurity investment
- regulatory capital planning
- critical infrastructure protection

ROI estimates compare:

- bounded, known mitigation costs
- against unbounded downside exposure
- under conditions of correlated failure

Quantification Approach

Orders of Magnitude, Not Point Estimates

Financial impacts are expressed in ranges (e.g., millions, tens of millions, billions) to:

- avoid false precision
- reflect variability across organizations
- acknowledge uncertainty inherent in systemic events

This approach aligns with best practice in catastrophe modeling and strategic risk analysis.

Treatment of Reputation and Trust

Reputational loss is modeled indirectly through:

- customer churn percentages
- deferred or lost future revenue
- market share erosion
- regulatory or contractual consequences

Trust degradation is treated as a multiplier not a standalone metric.

Limitations and Uncertainty

These scenarios intentionally do not attempt to:

- predict adversary behavior
- forecast timing or probability
- model exact blast radii or field strengths
- estimate total national economic loss

Those questions are inherently uncertain and, in many cases, unknowable in advance. The scenarios instead illustrate decision-relevant exposure.

Why This Methodology Is Appropriate

For risks with:

- catastrophic potential
- systemic coupling
- limited empirical data
- asymmetric downside

Waiting for certainty guarantees unpreparedness.

Scenario-based planning enables organizations to:

- identify fragile assumptions
- prioritize survivability
- preserve decision space
- reduce irreversible loss

This methodology reflects how responsible institutions plan for rare but consequential risks.

Supporting References

These references support the technical plausibility, dependency modeling, recovery timelines, and economic framing used throughout the scenarios and ROI analysis.

EMP and Electromagnetic Effects

- Commission to Assess the Threat to the United States from Electromagnetic Pulse (EMP) Attack, *Critical National Infrastructures* (2008)
- U.S. Department of Defense, *Starfish Prime Test Reports* (declassified)
- IEEE Transactions on Electromagnetic Compatibility (various)
- National Research Council, *Severe Space Weather Events—Understanding Societal and Economic Impacts* (2008)

Power Grid and Infrastructure Resilience

- U.S. Department of Energy, *Large Power Transformer Study*
- North American Electric Reliability Corporation (NERC), *High-Impact, Low-Frequency Events*
- GAO, *Electric Grid Resilience* reports

Disaster Economics and Recovery

- FEMA, *National Disaster Recovery Framework*
- Lloyd's of London, *Business Blackout* scenario analysis
- World Economic Forum, *Global Risks Reports*
- Insurance industry catastrophe modeling literature

Business Continuity and Risk Modeling

- COSO, *Enterprise Risk Management Framework*
- Basel stress testing methodologies (scenario modeling principles)
- DHS / CISA, *National Infrastructure Protection Plan*